On the Path of the LONE WOLF

Thoughts and Poetry of the Appalachian Trail

Stephen Lont

First Edition

PAGE PUBLISHING, INC.
New York, NY

First originally published by Page Publishing, Inc. 2018

Any and or all religious remarks or statements in this manuscript are specifically my own opinion.

ISBN 978-1-64298-073-8 (Paperback)
ISBN 978-1-64298-074-5 (Digital)

Printed in the United States of America

This book is dedicated to the loving memories of my mother (Carol Ruth) and father, especially to my father James C. Lont. By giving me his guidance, love, understanding, and total support, I was able to complete my dream.

Contents

Photos

1. Brother Tim, his wife Joyce, myself
2. Myself at approach to trail
3. Rocks on trail

4 through 5.
- Roan High Knob shelter
- Shenandoah trail
- AT map
- AT visitor center (Harper's Ferry)
- Myself on mountaintop

6. Night coming in
7. Cloud's in mountains
8. Mountains
9. AT
   - Shower
   - Jim and Molly Denton Shelter
   - Mosby's Raider's sign
11. Charlie's Bunion
12. Sunrise
    - Barn shelter
13. Doc and Myself.
14. Mt. Roger's, Virginia.
15. Uncle Johnny's
    - Manassas Gap shelter, Virginia
18. AT
    - AT
19. Mountains and Gaps

20. AT
22. Woodpecker tree
  - Cliff with road (Upper right)
23. Myself by stream in NC
  - Stoney Man Mountain
32. Glasgow Town (Manakin of Raquel Welch in the movie *10,000 BC*)
33. Rabbit
34-. John's Hollow signpost
35. Spy Rock
36. Deer in Virginia
37. Bear bags on pole
38. Night campfire
42. Devil's Racecourse
43. Timber Rattlesnakes protected sign
  - Cave
45. Cave in rocks
46. Path of rocks (Penn.)
47. Harper's Ferry
48. Harper's Ferry sign (Myself and Cousin Joyce)
49. Harper's Ferry mileage sign
50. Cup of blueberries
51. Yellow Springs sign
  - Tree overlapping rock
  - AT back country permit
54. My tent in pine woods
56. Barn shelter
57. Myself on mountaintop
58. My washing machine
59. Knife's Edge
60. Palmerton Police Station (PA)
64. Lehigh Gap (PA)
65. Rock stepping
67. Town below mountain
69. 1940s shelter (Bake Oven Knob)

71. Ladder going up rock (NJ)
- Waterfalls
73. Myself at falls
- Waterfalls
74. Church hostel
77. Church
78. New Jersey path
- New York-New Jersey state line
81. Flowers on path in AT
- "
- "
- "
- "
- "
- "
82. Sunfish Pond (NJ)
83. Crater Lake (NJ)
86. Bald Mountain Pond. (Maine)
90. Box Turtle
96. Sittin' Rocks
- "
- Rock and pack
- Sitting rock
98. White House Landing
99. Maine camp
102. GG and Bruno
103. Mt. Katahdin sign, and hiking stick

Why Poetry?

I decided to write these poems along the Appalachian Trail because after journaling for a while, it began to become a bit boring to me. I found it to be more interesting to write in a poem what a black bear was doing when I saw him than it was to record in a journal. (I saw black bear today.)

Acknowledgments

My thanks to my brother Tim, his wife Joyce, and my friend Tom Alderton, who encouraged me to publish this book.

My trail name was Lone Wolf. Here is a poem written about it.

The Wolf

The wolf alone wanders
Through lifetime, it seems.
Pondering always
On other one's dreams.

Though he loves his mate
And cares for her pups,
He finds that too late,
It isn't enough.
While he snarls and bites
And savors his kill,
In none of these things
The dream will fulfill.
He searches and roams.
His life, it may seem,
As he looks in the den
Of another one's dream.

I was given a ride from Michigan to Amicalola Falls State Park, Georgia, by a loving brother, Timothy, and his wonderful wife, Joyce. This was around Valentine's Day in February. There is a beautiful waterfall at the state park, and it is considered the approach trail to Springer Mountain which is the starting point of the Appalachian Trail. The falls is located nine miles southwest of Springer Mountain, Georgia.

Family

I have this brother and his wife who have stood by all my dreams.
They've always wished the best for me. They're family, indeed.
I thank my God most every day for blessings I've received.

A Walk on the Waters

I took a walk one day to ease the oceans of my mind. Some seas upon those oceans were of the troubled kind. At times I've felt like a piece of wood . . . just a-bobbing on those seas.

I knew I had to let that wood adrift . . . and let it be. When anchor sought, I found a place. A place strong and secure. No matter how high those billows, I knew I would endure. The highest ocean swell couldn't toss me about that sea. I'd let that piece of wood adrift. Let go and let it be.

Morning Blessing

Morning arrives. The air so fresh, so new. Grasses sparkling under clean, sweet dew. My heart sings out! It's great to be alive! Bless the Lord, oh my soul, and all that is within me. Bless his holy name.

Almost everyone who hikes the AT has a trail name. A trail name is a name other than their original first or last name that they are named because there is a specific reason for that name. For example, GG means Got Gumption. All trail names are in capital letters.

Trail Names

If you want to walk with *Lone Wolf*, you've got to walk *Downhill.* If you want to walk with *One-stick*, you don't have to be a *Hardcore*, *Catch-up* kind of a guy doin' the *Mud Flap* to walk *Uphill.* No need to be a *Gangsta* to know you're a *Slowpoke* if you've had *One Too Many.* When you're with *Nightwalker*, you'll see either *Goonies* in the dark or *Glow Worm*. If you hang with *McBride, Red Rocket, or Leif-E,* you might run into a *Pirate* or a *Black Cat. Braid* does the hikers' anthem for all of us to hear, and plays a mean harmonica, a beauty to the ear. *Captain Max* is a veteran a man for all to know, and sets a pace quite nicely, for that does surely show. A *Lunatic* can't be *Primitive* with *Divas* or *Babes In Boots*. Listen to *Radio Freak*, and don't *Yak* 'cause he's a *Jukebox Hero* who might just come *Unplugged. Work-Stay* can save his Jolly Rancher, saltine, and pickled egg for later.

To stay on the *Fast Track, Leadfoot* doesn't need a *Splinter* or he'll be headin' down some *Lonesome Roads.*

Scout doesn't need *Zen* or *Radar* to keep him on the trail. Well, that's it for this drop. Till next time and more mail.

Roan High Knob Shelter

(Near Carver's Gap, Tennessee)

This is the highest shelter on the Appalachian Trail at (6,275 ft.). At one time in the 1940s, this had been an old fire warden's cabin.

Appalachian Trail – Shenandoah National Park

The Appalachian Trail Conservancy

www.appalahciantrail.ogr/imagesa/mps/appalachian-trail-map.jpg

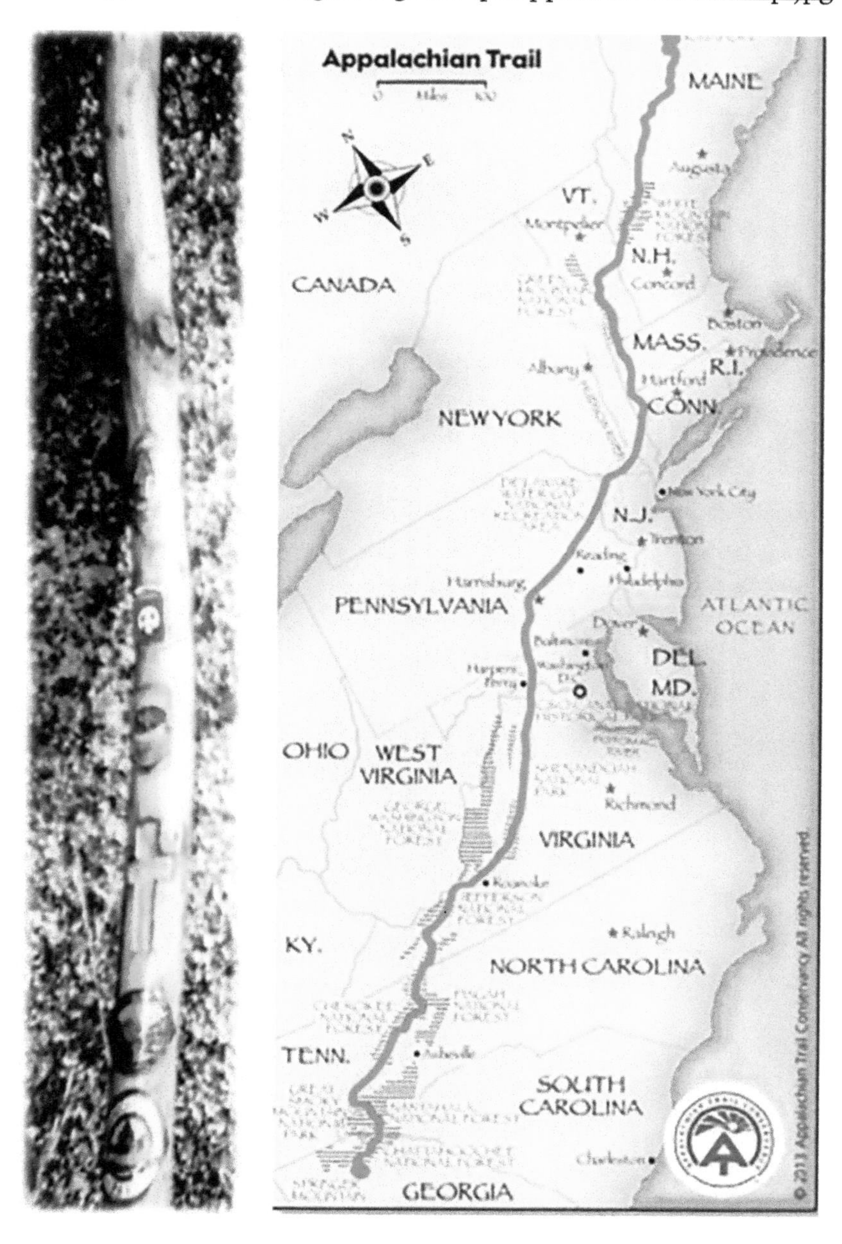

Appalachian Trail Visitor Center

(Harper's Ferry, West Virginia)

Time Alone

Everyone needs some time alone; a time to think you see . . . A time to be yourself in your reality. A time to be more consious of place's… that you are, of shallow thought's, and more deep, of place's near, or far,

A place to be more thoughtful of many things around . . . so let your mind just wander to all. Let go . . . abound.

Sassafras Tea

I took the root of sassafras and made myself some tea. When we were young, Mom taught us how to make that tea. I put the tea into my mouth, it savored oh so well!

When little boys, we'd figure, Ah! That's really swell!

Night coming in

Clouds forming in mountains

One Night in Georgia

I'm up on this mountaintop, you see, with nothing around. Very cold, cold, indeed.

The hail came down hard followed by sleet, then wind started blowing. 'Twas nothing too sweet.

My tent was put up in five minutes, you know, then man alive! The wind really did blow.

I went in my tent to try sleeping for real, but . . . the wind kept a-howling, and . . . trees started to go.

I said, "O dear Lord! Keep me safe, won't you, please? Keep me safe, O dear Lord, from these crashing down trees."

Smokies Unreal

There's beauty in the Smokies and not just in its clouds.
Some beauty in the Smokies lies just beneath the ground.
There's gold and micah and gems, you know.
The Smokies is unreal. So go and check those Smokies out and see just how you feel.

Déjà Vu

Sometimes I get that feeling . . .
I've been down this path before.
The mind plays tricks with me of
Some places that I've been.
I've felt that I have walked this part,
And
Walked this part again.
It's not really hard to
Figure; some parts just look the same.
So
Let the mind play tricks on you,
Pretend . . .
It's all a game.

One early evening, there were about eight or nine of us tenting in a meadow and this is what occurred.

The Owls

The Barred owls were out late last night, teaching their young, teaching in flight. As all parents will teach their young, this one's training had begun. As our guardians had trained us to eat, this one's parent to look . . . for meat. While they worked the forest section by section, most of the animals tried to avoid detection.

They hooted and howled very much indeed, and a few of the animals would not heed. A rabbit, they scared into bounding that night, and came down upon him while still in flight. Coming rapidly upon him, he screeched and cried. Poor rabbit . . . Poor rabbit . . . He could not hide.

Jim and Molly Denton Shelter had a solar powered shower!
(Virginia)

Jim and Molly Denton Shelter

Mosby's Raiders
(Northern Virginia)

This is a poem written about a rock outcropping in the Smokey Mountains that was named after a bunion on a man's foot. The man's name was Charlie Conner.

Charlie's Bunion

Charlie's bunion was sore, Charlie's bunion was sore, so . . . we climbed on it, climbed on it, climbed on it some more. Charlie's bunion was never more sore to the touch after . . . climbing all over his bunion so much.

In the morning, sometimes the sun can come up so quickly that it is there before you know it.

Sunrise

Sunrise steals into the sky, like . . .
A silent thief.
Hardly knowing it's lights are on,
revealing
the jewels of the day.

A barn, a shelter

Survival

It was cold that night, with snow on the ground. You know, the shelter was full. Doc and I slept under the shelter that night upon the frozen ground. In dried out leaves of dust and dirt, we were warmer than those up top. Much warmer than those up top, you know, so . . . Us, it did not hurt.

Doc and I

A note about Doc:

At first, I thought Doc's trail name stood for *doctor*. After asking him about it, he told me that it stood for *dude on couch*.

Mt. Rogers is Virginia's highest peak (5,729 ft.). There were many wild horses or ponies that live and roam freely on and around this summit. Every so often, the government has to glean or reduce the herd size so there is no danger of overpopulation or overgrazing of the herd.

The Horses

The horses at Mt. Rogers really aren't afraid. Why, I came around a corner and there was one that stayed! I was just two feet away, I had nowhere else to go, and the horse just stared right at me. 'Twas a brave, brave pinto. It was a brown and white pinto. Its coloring, quite attractive; hadn't seen a horse so proud, as long as I shall live.

The horses at Mt. Rogers really aren't afraid. Why, I came around a corner and there was one that stayed!

In Depth

If you want to go in depth, says I. We'll go in depth, you see. We all . . . in this world are lookin' for reality. Man gets so bored so easily with common things and stuff. Says I . . . look to God for answers, and life won't be so rough. Our Bible has the answers, God wrote for us to see. All common things and stuff, in man's reality.

Uncle Johnny's (Outfitter) in Erwin, Tennessee

Keffer Oak

I've never seen a tree so big; majestic in its size. Eighteen
feet thick, not merely a stick. Its limbs just rise and rise.

Three hundred years old now, all I can say is . . .

Wow!

How lumbermen forgot this oak is really
hard to tell. But royal beauty says,
It's just
So good . . .
And well.

Manassas Gap Shelter

Northern Virginia

I was a little leery of staying overnight in this shelter because I had read in the logbook that was in the shelter of reported sightings of black snakes and cottonmouth snakes in the rafters of the shelter. I stayed the night without any incident.

The Ballad of Ketchup and the Coyotes

They followed my friend one night, you see. He stumbled upon them. There were two, maybe three. He was hiking at night, when to his surprise, came over some rocks, and there they did lie. He'd hiked with his light on but much fog did abound, so 'twas hard to see them being all around.

They'd follow nearby him with glow in their eyes, and would not disperse till 'twas shelter nearby. He had hiked with his stick and nothing to use. Felt if they had attacked him, he had . . . nothing . . . to lose.

Psalm 23

Even though I walk through the valley of the shadow of death, I fear no evil, for your rod and your staff, they comfort me.

The Path Chosen

My feet have chosen paths of good and often paths of bad. The path of life, the path of death, not seeing which the other had. My feet now chose the path of life, assured of what they have. Step fresh with vigor, where once had weakly trod.

The Appalachian Trail consists mainly of a long series of mountain ranges, depending on how you look at it from south to north or north to south on the eastern coast of the United States. In the southern states, the low spots or valleys are called gaps. In the northern states, the low spots or valleys are called notches. This poem gives you a feeling of what it is like to hike them over any length of time.

Mountains and Gaps

Up the hill, down the hill, up the hill, down the hill, you don't look up, you just look down,
For . . .
Just more hill you'll see.
You don't look up, for all you'll see is hill eternally,
And . . .
Gaps are funny things, you know, for . . . after gap, it's just more hill, you see.
I used to get excited when gap I'd see on map,
Knowing . . .
Downhill was for sure,
But . . .
in would sink reality.
For . . .
just another hill I'd see.

Thinkin' Time

I've been talkin' to myself, just a-walkin' down this path. It's not hard to figure out, just try to do the math. There's no one else to talk to but me and only me. Not him, nor her, nor they, nor them. Not very hard to see. There's no time to stop and rest, just time to stop and think some more. Just time like that, at best.

McAfee Knob was one of the highlights of my hike on the AT. I had a hard time explaining it, so I didn't really attempt to try. It is like trying to explain the Grand Canyon to someone who has never seen it before.

McAfee Knob

It's a blue-ribbon day at McAfee's Knob. The sky is clear for all.

You can see for miles around, and . . . beauty, surely . . . does abound.

The Woodpecker

Woodpeckers have a thousand beats; for them, the common cure. For pecking, it comes easily. A home, a nest secure. They eat the insects on the trail get rid of bad ones and the frail . . . and pecking comes so fast, you see, try counting them . . . you or me. I guess that's why they call him . . . the woodpecker, you see.

June Day

It's a hot day in June. It's a hot day in June,
and . . .
the comfort I have is the pack on my back,
and . . .
a cool, cool breeze through the trees. It's a hot day in June.

The gnat is one of those insects that could become very irritating on a hot summer day.

The Gnat

I dislike the gnat, and I'm sure of that. For him, being a bother, I'll take off my hat. He'll fly in your eyes when it's hot, you know, and don't open your mouth 'cause in he will go!

He's a pain to get out of your eyes, for sure. So, do heavy the bug spray, for that is the cure.

I dislike the gnat, and I'm sure of that. For him, being a bother, I'll take off my hat.

As I have mentioned, there are some things I just don't like.

Things That Creep and Crawl

I do not like some things, some things that creep and crawl.
I do not like the centipede.
I do not like the Pygmy rattler.
I do not like the spiders.
I do not like the cottonmouth.
I do not like the gypsy moth.
I do not like the blacksnake.
I do not like the caterpillar.
I do not like the ring-necked snake.
I do not like the coral snake.
I do not like the millipede.
I do not like the gnat.
I do not like the rattlesnake.
I do not like the snake.
I do not like some things, some things that creep and crawl.

Whip-Poor-Will

Whip-poor-will, whip-poor-will, please be still. It's too early today. It's three-thirty in the morn that the whip-poor-will starts on his musical horn. Why does he have to get up so soon to blow? He could certainly wait on that horn.

Don't you know?

Whip-poor-will, whip-poor-will, won't you please be still? Whip-poor-will, whip-poor-will, please be still.

Backache

I had this major backache. It would not go away.
Sometimes you just get stuck, get stuck along the way.
My body was so weary.
It was time to stay and pray.
I do not pray, just in a jam. It's not the thing to do.
I talk to Father every day,
For . . .
He'll watch over you.

Some areas will have signs as you approach them. The signs make you aware of the mileage or the name of the area you are approaching.

The Sign

The sign said Tinker Cliffs. It said the sign. Oh sure, says I. I'd like to know what Tinker knows, before . . . the cliffs, go I. The pots and pans the tinkers fixed had value, that's for sure. When cliffs I make, most likely make . . . I'm sure, I will endure.

Sly Bird

I saw an injured partridge along the path today.
It's wing was clearly broken, of that I'd have to say.
It made some frantic noises as it scurried from my reach. I thought that I might mend it's wing, and let it free again.
How wrong! How wrong did I find, when much to my surprise . . .
It took up wings and flew right off before my very eyes.
I pondered for a moment and figured it all out.
That mother Pat was fooling me to draw away from young ones, see?

When the colonial army fought the Revolutionary War against Great Britain, this was an area where the tinkers, the men who fixed pots and pans for our colonists hid. I imagine that they also poured lead into molds for musket balls here. This is an area where my imagination ran wild.

I, myself, could not envision the British soldiers marching the long distance to these cliffs in formation uphill.

Below Tinker Cliffs

We won the war in 1776, in part to tinkers on those cliffs. I've spent some time there down below. It's a place the British feared to go. A place for tinkers fixing pots, and spring fed water lots and lots. A place of hideouts, a spot to live. Easily defended, nothing to give. The British wouldn't march so far . . . to Tinker Cliffs uphill. They'd rather be in comfort at home with Mom and Lil. We won the war in 1776 . . . In part to tinkers on those cliffs.

When you've been out in the woods a month or so, you can easily lose track of what goes on in the rest of the world.

Unaware

Sometimes you lose track of . . . goings on out there.

Of news and politics, you're totally unaware. You'll lose track of time, sometimes you'll skip a day. But . . . you'll make it up, you see, in every other way.

There are some towns that you enjoyed spending a little bit of time in for one reason or another. For me, this was one such town.

Glasgow Town

Glasgow had a nickname. The town that time forgot. I think the town of Glasgow can offer quite a lot.

It's peaceful and serene, just nestled in the hills.

If you like the peace and quiet, then Glasgow has no frills. It's gorgeous and so quiet, just sittin' right out there. That little town of Glasgow, so good and quiet and rare.

An example of my quick jotted-down journaling:

THE RABBIT.

THE RABBIT,
HE PAUSED,
HE STOPPED THERE,
HE LOOKED AT ME.

DO I WANT RABBIT, FOR
SUPPER, SAY'S I?....
NO, SAY'S I.

THE RABBIT, HE PASED,
HE STOPPED THERE, HE LOOKED
AT ME, THEN......
SLOWLY... HOPPED AWAY.

The Rabbit

The rabbit,
He paused.
He stopped there. He looked at me.
"Do I want rabbit for supper?" says I.
"No," says I.
The rabbit,
He paused.
He stopped there. He looked at me.
Then,
Slowly . . .
Hopped away.

The Storm

You could hear it
coming from beyond . . .
coming from a distance.

The first thing to hit,
coming from beyond . . .
coming from a distance was . . .
the wind.

The wind grew steadily
from beyond,
from a distance,
at John's Hollow,
until . . .

The wind was upon us from beyond, then . . .
In a thundering downpour as a constant drumbeat of
things to come, came . . . the rain.

The rain came steadily,
At . . .
John's Hollow throughout the night.
Tapping drumbeats on my tent
At John's Hollow.

APPALACHIAN TRAIL
JOHNS HOLLOW
61 / 147

Spy Mountain, or Spy Rock as it is called, was a high rock with a good view all around the surrounding countryside. It was rumored that the Confederate Army used this high point to watch the Union Army's troop movements.

Spy Rock

When the Confederate Army wanted a place that they could see, it was the rock, Spy Rock, the place they chose to be!

They could see for miles around! Their enemy, the Union, was easily found. They could watch them march through a gap.

They could watch their campfires as they lay at their nap.

When the Confederate Army wanted a place that they could see, it was the rock! Spy Rock! The place they chose to be.

This sounds like an advertisement for tourism in the state of Virginia. I saw twelve deer in one night on Spy Mountain at 7:20 p.m. They were more curious than afraid of me.

The Deer

The deer are here in Virginia. I've seen more deer in Virginia than I've seen in any other state.

If you like to see deer, come here. Come to Virginia.

The deer are here in Virginia.

A manufactured pole on which to hang your bear bag (food).

Here are two poems written about the same event.

The previous night, I had stayed at Maupin Field Shelter about seven miles before I had made it as far as a place in Virginia called Dripping Rock because this was a place where a spring came out of a rock. I continued on until running into Brooke Osborne. (Wampus Cat) Wampus had already set her tent up, as it was beginning to rain pretty steadily, so I did the same.

It wasn't a regular campsite, so we set up a stealth campsite. A man had come by on a mountain bike and warned us of a lot of bear activity in the area. Shortly after setting up my tent, others began to join us. Here's what happened after we had all gone to sleep.

Bushman's Treat

I hung up my bag off the limb, don't you know? And the Bushman says, "Hey! How 'bout two in a row?"

I said "Fine, sure fine, just fine" really slow.

I don't care, I don't mind 'bout two bags in a row. But . . .

The bear, he came by. He came by, sure, alright, and stole Bushman's bag, just a meal in the night.

Wampus and the Bear

It was two in the morning when Wampus awoke to something out moving, making noise, so she spoke.

She said, "Hey! Who's out there? Why leave at night?" 'Twas the bear for his bag, 'twas the bear in the night. We turned on our headlamps to find 'twas no joke. For the bear had absconded, trying not to be rude, and had totally left with the bag and the food.

Home Fires

At night, some of us . . . some of us who aren't in quite as
big a hurry, we pause. We like to keep the
home fires burning.
We pause and . . . take time out from our journey.
We pause in life . . . to think of those at home,
those loved ones in life's churning flurry.
At night,
some of us,
we pause . . .
We take time out,
And . . .
Keep the home fires burning.

I saw this strange occurrence in the Shenandoah National Forest near Little Roundtop Mountain in the state of Virginia. The deer had a fawn on the ground nearby.

The Standoff

The bear huffed, and the deer stood still.
The bear huffed and huffed some more . . . and the deer stayed.
As the bear huffed and the deer stood still . . .
I slowly, so slowly, walked away.

The Soft, Gentle Touch

The wind plays softly through
the trees this morn.
Softly . . . softly . . .
such a gentle breeze.
The wind plays gently
in the trees this morn.
Gently . . . gently . . .
A rhythm in the trees. I can hear the breeze,
as a gentle rush . . . Very softly, she plays
with . . .
a soft gentle touch.
The wind plays softly in the trees this morn,
Softly . . . softly . . .
Such a gentle touch.

Ear Music

Gentle breeze upon the pines tonight. The rain falls softly, too.
The fire is flickering, oh, so bright, with music in my ears.
The sounds are lulling me to sleep . . . so softly in my ears.

There's this stretch of the trail in Pennsylvania that is a rough area. This part of the trail is referred to as the Devil's Racecourse.

The Devil's Racecourse

The devil's got this track. It's hell on wheels, you see. It's called the Devil's Racecourse, and that's reality. It's made of rocks and boulders . . . 'tis nothing fun to race, so watch your step! Don't slip and fall.

You'll fall right on your face, it's really not too funny. The devil broke a horn. So if you think it's funny, just go ahead and scorn.

There were all kinds of snakes on the trail. The most snakes that I spotted were in the state of Pennsylvania. Most of the poisonous snakes live in areas where the temperature is warm the longest part of the year. My general feeling is that I don't like snakes, and I would rather see them dead. But some snakes are protected by law. The Timber Rattlesnake is one such snake.

The reason that this snake is protected is because it only breeds a couple of times in its lifetime, and when it does breed, there are only a few of the young snakes that do survive. If everyone was out killing rattlesnakes, they would soon become extinct.

The Snake

The snake is but the devil's own. A piece of hell, you see.
Black snake . . .
Rattlesnake . . .
Cottonmouth . . .
All.
For he was but a creature, till tempting Eve to fall.
He's such an evil creature, his head I'd love to crush. But there's plenty of snakes out on that trail . . . For me, there is no rush.

Once in a while, you will have a day where you can't hike or just don't want to for some reason. This is commonly called a zero day.

Zero Day—Nothing Going On

It rained all day. Too hard to do much hiking.
There is nothing going on, nothing to my liking.
Rain gear, I did not don. Cloud's coming over
mountaintop. Hot, humid, sticky day.
I placed my tent in hollow down, away from wind
on mountaintop. Hot, humid, sticky day.
Thick fog on mountaintop.
In clouds, I know I'll be till morning and I'll see.
Hot, humid, sticky day.
Nothing to my liking.
It rained all day.

The Cave

The cave in the rocks.
You do not know
what lives
In the cave, in the rocks.
It could be a home . . .
for snakes, you see.
Or . . .
A den for a bear with cubs,
two or three.
The cave in the rocks,
It . . .
intimidates me.

The Devil's Plan

The devil's up all night, you know, sharpening boulders—those boulders far down below. He gets a lot of joy in this; it makes his snake just want to hiss, so . . . Watch your step along the path and don't fall in a hole. Be sure to use your hiking pole 'cause the devil wants your very soul.

John Brown tried to help the slaves rise up against their masters, and his rebellion failed. He was a white man from Kansas. Many people believe that this uprising is what helped spur on the Civil War between the north and the south. Where this occurrence took place was in the state of West Virginia.

Ole John Brown

Harper's Ferry was made famous by Ole John Brown, you see . . . for freedom's all he wanted for every slave that be. So Ole John Brown attempted to seize the armory, but sadly, Ole John Brown was crushed by Colonel Robert E. Lee.

HARPERS FERRY

Harpers Ferry
National Historical Park
Appalachian National Scenic Trail
← Maine
1,165 mi
Georgia →
1,013 mi

The Altar

I built an altar to God one day. 'Twas nothing simple, as you might say.

I built the altar on mountaintop of rocks and things and stones, whatnot.

I built a fire of twigs and wood and prayed "Dear Lord" to make . . . me good. God answers prayers to you and me . . . not always the way we want them to be.

God answered my prayer, 'tis sure to see. For with Him, I'll be eternally.

This poem was written after a man with the trail name of Pyro walked by me. I let him pass me by on the path because he was talking loudly and had earphones on. It became obvious to me that he was taking voice lessons and wasn't really paying attention to what was happening around him.

The Cub

I'm coming 'round this corner when all kinds of noise I hear. Off to my left, there's a tiny cub . . . just a-coming down this tree. How neat! How cool to see . . . that tiny cub just a-coming down that tree.

I had let Pyro go by me about ten minutes earlier. What I think happened is, the loud talking made the mother bear scare the cub up the tree. When I came by ten minutes later, the cub was coming down the tree. I didn't see the mother bear around, so I kept going, not stopping to watch the cub. If mama bear came back by, I would have been in trouble.

This happened by Table Rock, Pennsylvania.

The Very Berry Cure

I'm walking on the path today, what would you think I'd see? I look to my right, I look to my left, and there's nothing but blueberries. So, I stopped to take a break, you see, to fill up my Jetboil cup. It was time well spent for a break. I thought to pick some blueberries.

Those blueberries were great, you see.
Well worth the time I'd spent.
For . . .
On the trail, it's one sure way
To . . .
get your antioxidant.

The Ghosts of Yellow Springs

In the 1850s, men would come from all around. The men would come to work the ground. It was the coal! The coal! The coal they had found. The women would come and follow the men. Those womenfolk would bear children. The village of Yellow Springs was formed as a busy little town all for the simple reason—there was coal down in the ground. Yes, there was coal, coal, all around. The men are long since gone now, women and children, too. Families are a thing of the past, for there's nothing left to do. There's nothing left of Yellow Springs but piles of rocks to see.

Yellow Springs has vanished along with the memory.

The ghosts are all that's left now. The ghosts of things long past. Yellow Springs is but a memory of things that weren't to last. All that's left is the family plot . . . for Yellow Springs has gone to rot.

Of Rocks and Things

Look at the rock, or should I say boulder?

It's a very big rock, a very big boulder.

The rock has been here . . . six hundred and fifty million years or so, give or take a day or two.

Look at the plant, or should I say tree? It's a very big plant, a very big tree.

The tree grows over the rock; the plant overlapping the boulder.

The tree has been here . . . one hundred and fifty years or so, give or take a day or two. Look at the rock. Look at the tree. Neither can reason like you or me. We might live to be . . . who knows?

In God's hands are all that grows.

Form 10-404

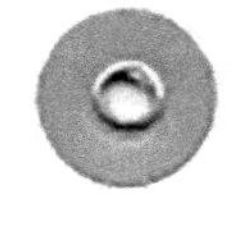

NO OPEN FIRES

U.S. DEPARTMENT OF THE INTERIOR

NATIONAL PARK SERVICE

BACKCOUNTRY USE PERMIT

The visitor must have this permit during the visit.

When signed, this single-visit permit authorizes:

NAME STEPHEN LANT

ADDRESS 1130 W 27TH ST

CITY HOLLAND MI 49423

To visit SUP

Give best estimate of start and finish dates

Location of entry ROCK HARBOR

Location of exit

Primary method of travel WALK

Number of people in group 1

Number of pack or saddle stock 0

Number of watercraft or other craft 0

CAMPSITE NAME

VARIOUS

DATE

PERMITTEE'S SIGNATURE

DATE

ISSUING OFFICER'S SIGNATURE

FASTEN THIS TAG TO YOUR PACK, SADDLE, BOAT OR TENT

The main thing that attracted a black bear's attention was food or the smell of it. The best thing was to put your food in a bag called a bear bag) and hang it in a tree. We were told to hang this bag at least ten feet up in the air and six feet or more out on a tree limb.

The Bear

The bear was out last night. I could hear him walking, looking for food.

A bear was out last night. He came near my tent, looking for food.

A bear was out last night. On my harmonica, I gave him a tune.

A bear was out last night. That sent him off in fright, causing all to be alright. For me, a great delight.

A bear was out last night.

Heaven and the Pines

My human idea of heaven is . . . a campsite in the pines.
If heaven could be a bit like that, to me, it'd be just fine.
I love the smell of pine so much, it smells so fresh and clean
I love to sleep at night upon soft needles of evergreen.

A Chipmunk's Life

At Phillip's Canyon Spring, I sat down to take a break. This chipmunk hops up on my shoe, he looked at me as if to say, "Hey! Hey! Who are you?" He jogged around a little bit, then ran into the wood.

I wish that he could speak to me. I wish he really could.

This is a poem of a man who hiked with us from Georgia, to the state of Maine. His trail name was the Crow.

The Crow

The crow is out here with us. He walks with us, you see.
He cannot hear or read your lips, he writes everything that can be.
The crow is out here with us. He walks with us, you see.

This was a mountain or hill in Pennsylvania that was, well, read it for yourself.

Steep Descent

I'm coming in Port Clinton
on
this non-friendly hiker hill.
It
reminds me of the song, the
song of
Jack and Jill. How Jack fell down and broke his crown,
and Jill came tumbling after.

One day I was near a town called Port Clinton that had a river running past it. All I know is that some of my clothing had begun to gain in a sweet smell, so . . .

Wash Day

I took my stuff down to the creek and got my camp soap out.
It was time to do some laundry, of that there was no doubt.
I tossed my quarter in the creek, and said, "Here, that's from me."
I felt much better after all, when all my stuff was clean.

The Knife's Edge

There's this place called the Knife's Edge. It's razor-sharp, you see.

You don't want to fall on Knife's Edge, or a casualty you'll be.

It's not so easy hiking after raining on the ground, besides the fact and pointing out, that rattlesnakes abound, So be careful there on Knife's Edge.

Be sure to watch yourself . . .

You don't need to be a casualty upon some rocky shelf.

Lockup

I'm staying at this lockup in Palmerton, you know. The police in Palmerton put on quite a show. They're generous and good. They give hot showers, too.

So . . . if you stop in Palmerton, be sure that's what you do.

The mountains around Palmerton, PA, have been scarred by one hundred years of mining for the metal zinc. The E.P.A. stopped the smelting of zinc around 1980.

Just as oil drilling can be destructive to nature as we have seen in the Gulf of Mexico, man has also ruined some otherwise beautiful parts of nature by mining the land, as we have seen with coal mining in other parts of our country. Here is another type of mining that I saw on my hike. Very destructive to nature but beneficial to mankind. We need the oil, coal, and zinc that our land provides. Ladies, zinc is commonly used in most of your eyelash make-up.

To anyone who is a denture wearer, zinc is a common ingredient in the Fixodent that is applied to your denture before it is put back in your mouth.

Denuded Healing

Those mountains around Palmerton have zinc in them, you know.

By man, they've been denuded. They look ugly. They look cold.

Man has to make a living, and the zinc provided that. But it left a scar on nature, and the healing doesn't come fast.

The government set up a fund called Super, wouldn't you know?
They reclaim the land,
Reforest it,
And
Try to help it grow.
Man always tries his best, but still leaves scars behind.

This was an incident that happened to me near a place called Leroy Smith Shelter around 7:20 one evening.

A Hungry Bear's Life

I walked up on a bear last night. I stopped, thought, *Hey! What's that?*

His back was to me. All I saw was fur; a lot of black.

I stood there for a moment to think what next to do. I reached down for my camera and remembered it was packed.

So . . .

I started very slowly to retrace my steps back.

He must have heard me moving, for turning 'round, saw me.

He must have been so startled. He ran off! Ran off, you see.

He looked to be quite young, about a year and one-half old. If he had been much older, he might have been more bold.

I walked my way up the path to see what he had done.

He'd been turning over boulders to find what he could . . . of grubs and ants and larvae.

To him, it was so good.

Keeping the Bear's at Bay

Things that people say to keep the bears at bay:

Wampus Cat says, "Hey! Who's out there? Why are you leaving at night?" Doesn't work. The bear has already absconded with the food.

Growler says (while sleeping under a tarp after hearing a loud noise at night), "Hey! Don't you know there're humans out here?" Seems to have worked, Growler is still with us.

Man on Kittiny Mountain (to two young bears), repeats out loud over and over, "You go your way, I'll go mine. You go your way, I'll go mine." Worked. The bears came close, then ran away.

Lone Wolf's way. He says loudly, "Hey Yogi! Let's go get some picnic baskets at Jellystone Park."

Works every time! The bears leave in a big hurry, trying to find the picnic baskets.

Slack Packin'

While in Pennsylvania, I did slack packin', you see.
I was stayin' at my cousins; my cousins in Carlie.
The slack pack is a way to pack lighter for the day.
It's really very nice, for you can put on miles that way.

With a lighter pack, you can tighten up the slack. All you need is food and water. Food and water for the day.

There were some days that you could spend all day stepping on or around rocks or tree roots. It could become very tiring after spending a day like this.

Rock Stepping

My feet are tired of rock stepping and rock stepping so much.
All my feet desire is . . . the tenderest of touch.
If feet had brains, they'd surely try avoiding rocks, I know . . .
But feet can't think and merely step in trail and go with the flow.

While in Pennsylvania, I passed through an old ghost town from the 1850s. At one time, the town had been a prosperous coal mining town. Near the edge of the town was an ancient cemetery. Here are a few of the markers on the headstones of the graves:

In memory of John Proud
Born in Durham England
Died May 1854.
Age 52 years and 16 days

Affliction sore, long time I bore. All human skill was vain, till God did please to give me cure, and free me from my pain.

CATHARINE
Daughter of John and Elizabeth Blackwood
Died June 16, 1854
Age 11 years, 1 month, and 7 days
In memory of the late Andrew Alleno
A native of England, who met his death
by accident at Gold Mine Gapo
June 9, 1854
Age 30 years, 2 months and 17 days

Here lies beneath this humble soil
The noblest part of nature's God.
I heard, once worn with gratitude, with strength,
and courage was endured. J. Allen.
Few hearts like his with virtue around.
Few hearts with knowledge so informed.
If there another world, He lives in bliss.
If there is none, he made his best in this.
B. Burns.

One day I was sitting on top of a mountain just looking down on a small mountain town, and fleetingly wondering just what kind of sinful nature was going on down below me.

Below the Mountain

I'm sittin' on this mountaintop, just lookin' . . . at the world. The complicated things in life, so easily come unfurled. If God had made a life of strife, we all would surely know. But sinful man chose life of strife and woe below . . . oh woe.

Tin Cans in the Sky

I'm out here in this wilderness, watching tin cans in the sky. Those jet's, they go so speedily, while step by step go I. Technology, so awesome. You know it to be true . . . has passed us by so quickly, and thrown us all askew.

July Day

It's a beautiful July day. Not too hot, not too cold.
There's beauty to behold.
It's a beautiful July day.

The Heron

The heron stands upon one leg, overlooks his water domain. He scans his waters quick and quiet. His scan is not in vain.

He spots his fish in waters by, and quickly probes with bill.

Our heron is rewarded with another tasty kill. His head goes back and swallows catch, his catch, in one smooth show.

He then moves on to find more dinner, dinner deep below.

The Window

There's this window of the mind
That . . .
God opens up for me.
Of
all the things I love in nature that I see.

The Flip-Flop

I'd enjoyed my hike so much, I was running short on time.
I was running short on time to meet my fall deadline.
The weather might get bad, the time would be to tight, so
I decided that I'd flip and do a flop that night. It's kinda like a pancake. When one side is so well done, you flip that pancake over and the other side's begun.
I stopped in New York state and hopped a bus to Maine to continue my ole journey for eight hundred miles remained.

The Waterfalls

The waterfalls are here . . . tis golden on the ear.
Swiftly since times of old, the falls go tumbling by.
Watching the tumbling waters . . . the mind
is absorbed and pulled in by them.
As thoughts go rushing by . . . the eyes are
absorbed and pulled into them.
Swiftly since times of old . . . the falls go tumbling by.
The falls rush by . . . such a beauty to behold.
To behold . . .
To behold . . .
To hold . . .
To hold . . .
Old . . . old . . .
Old.

The Church Hostel

The hiker's friend on the AT, you know, are
the church hostels. The church hostels
that are
Where you want to be.
They . . .
ask but a donation, and don't force belief on you.
They . . .
lead by their example, and let you choose just what to do.
Their . . .
charges aren't outrageous. They're very minimal.
They're . . .
your friend along the path, as Jesus Christ would show.

Acceptance

Until you and nature see eye-to-eye, you'll never be happy hiking in a consistent, steady, downpouring thunderstorm or rain that lasts all day.

I'm not sure, but I think that this bird was attracted to a shiny fish hook that I wore in my hat.

The Grey Jay

The grey jay, he came bobbing. He came bobbing 'round my head.

He would not go away. He'd rather tease, instead.

He had no nest nearby to be protective of his young.

He enjoyed his little game and would not go away.

For him to stop his game, for him to go away, I had to get myself up. I had to leave his play.

The Issue

As I walk the path today,
There's an issue that I see.
It's an issue about faith,
Of . . .
Who'll watch over me.
The song comes by my mind.
The one I'd always sing.
Trust and obey, for there's no other way
To . . .
Be happy in Jesus,
But to
Trust and obey.

The Jersey Walk

One of the most enjoyable states I've hiked
was New Jersey, that's one I've liked.
The glaciers smoothed off mountaintops
as if you're walking on sidewalks.
There's never any lull, the views are all so beautiful.
So . . .
When Jersey invites me back someday,
it's probably Jersey I'd like to stay.

New Jersey, New York state line

Peace on the Rock

A duck was there with ducklings upon that shallow rock in the midst of Stoney Brook with swallows gliding round. The ducklings rested peacefully with mother on that rock, while swallows kept on feeding around that peaceful rock (Stoney Brook, Connecticut).

Prelude

1. An introductory performance, action, event, etc. preparing for the principal or more important matter. 2. Music. a. A strain, section, or movement introducing the theme or chief subject, as of a fugue or suite; b. An opening voluntary in a service (Source).

Prelude to a Storm

The owl is flying 'round, just a hooting in the wind. Lightning shows the way in the clouds up overhead. Thunder keeps a-rolling in clouds just passing by, and the owl is flying 'round just a hooting in the wind. 11:15 p.m. Silver Hill Campsite. 11:30 p.m. Heavy downpouring rainfall.

The Flowers

The flowers on the trail have beauty that's hard to match.
They're so delicate and frail, much fragrance they unlatch.
The smell is very sensuous, almost sexual in design.
It'll stop you in your tracks as if
flirting with the mind.
Beauty so unbounding, thinking you're in Eden.
Step lightly on the path, be sure to be a-heedin'.
Nothing else hath,
as
the flowers on the path.

Sunfish Pond

Sunfish Pond was a pond in New Jersey that I saw.
It was caused by glacial action; it put my mind in awe.
Such beauty to behold . . . It's like nature's piece of gold.
I'm glad the sun was shining when Sunfish Pond, I saw.
For long, so long, many years ago, Sunfish Pond still had to thaw.

Bullfrog Serenade

I can see the leaves' soft movement in the fullness of the moon.
All is oh so quiet, I can hear the bullfrogs swoon.
They are looking for a mate to continue down the line.
The future generations, generations of their kind.
All is oh so quiet . . . Here at Crater Lake, for . . .
The bullfrogs are out searching, searching for a mate.

This is one town in the U.S.A. that had American flags lining the streets on a regular basis.

I'll never forget the hospitality of this small town in the state of New York.

The Mayor of Unionville

There is a place called Unionville, a patriotic town. The mayor of old Unionville won't let a hiker down. The mayor of old Unionville invites hikers to his home. The hikers like his company, the hikers will not roam.

The hikers do their laundry there.

The mayor's a man who likes to share.

He feeds them two hot meals a day, and lets them take a shower. After supper, watches a film with them of self-esteem and power.

The mayor of old Unionville won't let a hiker down. He's a patriotic man in a patriotic town.

The Racoon

The bandit, he'll steal from you. Steal, yeah, that's right! The bandit, he'll steal from you deep in the night.

Don't leave out no objects of silver or gold. Don't leave out your wristwatch 'cause . . . He'll want to hold.

The bandit is bold.

The bandit will hold.

The bandit will steal from you deep in the night. The bandit will steal from you. Steal! Yeah, that's right.

Postcard

It was postcard picture perfect on Bald Mountain Pond tonight.

The harvest moon was in, showing everything in sight.

The loons were out there feeding in shallows near the shore.

Some cirrus clouds lay past the moon, you could not ask for more.

A fish came up for mealtime upon that mirrored lake.

The beauty of it all was oh so much to take! It was postcard picture perfect on Bald Mountain Pond that night.

The harvest moon was out showing everything in sight.

Feelings

I don't know for sure what it feels like to be one, but today . . .
I took a bath in the creek.
I felt like a nudist today, I took off every stitch of clothes.
I felt like a nudist today,
My socks came off! My socks came off!
I felt like a nudist today,
My shirt came off! My shirt came off!
I felt like a nudist today.
And last, but not least, my underpants!
I felt like a *totally nude nudist* today.
I felt like a *nudist* today!

Shaw's Place

There is this home, you know. It's a special place, you see. It's called Shaw's. Yeah, Shaw's, that's what it'll be.

They'll call you there for breakfast, seven right on the dot. They'll feed you everything you need, and won't leave you for not.

He comes around the first time, saying "What you want to have?" Meaning . . . Will it be one? Or two? Or three? Or four? Of every course that be.

I started out with two flapjacks come there hot right off the griddle, see? Along with that, two eggs, the best that ever be. Then come the two pork sausages, and bacon strips, also. He tops it all off then with hash browns, you ought to know.

"How'll you have your eggs?" says he. "Over light," says I. They'd never heard of that in Main, it made me want to cry. Well, then, there comes the OJ, and there's hot coffee, too.

I'm done with my first helping, he comes back again.

He says "Two? Three? Or four?" "I'll have two," says I again.

The third time he came back and asked me once again, I said three that time should fill me up, would hold me now and then. Well, you know to top it off, it was only seven bucks.

There is this home in Monson, Maine. It's a special place, you see . . .

It's called Shaw's. It's Shaw's. It's Shaw's, that's what it'll be!

This was a poem that I really enjoyed writing, very true to form. As in some of the most northern states, I found that one day you could be in a very nice and sunny area, and the next thing you realize, you could find yourself walking in a deep, dark, forbidding, swampy area where your mind would begin to play tricks on you.

Winged Monkeys

I thought I saw winged monkeys out in the woods today.
I could have sworn I'd seen them, there was no other way.
Those woods can get so gloomy and very dark at times. There're tree roots oh so gnarly and branches thick with vines.
I felt like poor ole Dorothy with Tin Man, grayish blue, and cowardly Lion with Scarecrow, who didn't know what to do.
Those woods can change so quickly from bright and sunny spots to . . . Dark and strange and spooky where you can smell the rot.
I swore I saw winged monkeys out in those woods today.
I know I must have seen them, there was no other way.

Something that I noticed more than once after achieving some very high mountain peaks was that I began to have encounters with box turtles in the middle of the path. The more I encountered this the more I began to wonder just how many eons of generations of mating did it take for this box turtle to reach the summit of this mountain? As slow-moving as the turtle is, I can imagine it taking thousands of years.

The Box Turtle

The box turtle in the path, he looked at me as if to say, "What's the hurry?"

"I've got a schedule to keep," says I.

Slowly moving away, he said,

"So do I."

By . . .

And . . .

By.

Winners and Losers

There're some people out there on that path, the good ones and the bad. The good ones easily remembered the bad ones sure to fall. The winning ones fun to know, for in their faces you'll see a glow. It's a purpose, a drive . . . they all know. The bad ones, I'd just soon forget. The preacher, who stole Spiderman's boots, I'm sure of that, he'll soon regret. There're some who have nowhere, nowhere to live. They stay at the shelters, wanting someone to give.

You see, there're homeless out here, too, with nothing much better, much better to do. I wish it wasn't always so. Try to look into their eyes, look for that glow.

There're some people out there on that path . . . the good ones and the bad.

Sometimes you would walk in the rain all day with your rain gear on. Sometimes you didn't realize it was going to rain until it was too late. I didn't like hiking with my raincoat on unless I had to because it could get too warm. Because they were totally waterproof, your body could not breathe through the coat, causing the warmth to get trapped inside. You had your good points and bad. It was a good thing to wear on a cold night inside of your sleeping bag.

Rain Day

It rained cats and dogs today. It rained like cats and dogs. If I hadn't known much better, I'd have felt I was part frog. I put my cover on my pack, and then my raincoat, too. By then, it was too late to keep them dry, my shoes. I walked in the rain, the pouring rain, and puddles went straight through. For wet is wet, and dry is dry, too late for them ole shoes.

After a couple of experiences like this, I learned to put bags over my feet to keep my socks dry. Bags could also keep your feet warmer if it got colder at night.

The Honeybee Tree

The bee tree is out there, it's plain for all to see.

The bee tree is out there, just for you and just for me. I change my mind so quickly, to . . . let go and just beware, and let the bee tree stay out there. Stay out for just the bear.

Of Bears and Mice and Men

I started my hike just a-stompin' my pole out in the woods. I started my hike stompin' to keep the bears at bay. To me, it was so simple . . . there was no other way. You see, I'd been born and raised afraid of bears, I guess. But the bears are just as afraid of us; afraid of us, at best.

The black bears are so timid; afraid of us, you see. The only ones to be afraid of are the fed ones that they be.

To bears, it's learned behavior. They're out living in the woods. Once he's learned you'll leave your food around, now that's not very good.

The only time they're aggressive is if they're always fed. As the saying goes, a fed bear is really good as dead.

The only other time to be scared of them, you know, is if they have their cubs around. Then stay your distance . . . Woe!

And then I'd sleep in shelters just a-yellin' at those mice. You know it wasn't normal, it really wasn't right.

So . . .

I see the humor of stompin' in the woods, and now screaming at those mice of which I never should.

Baxter was the name of a moose in the state of Maine that was named after the state park in which Mt. Katahdin, Maine's tallest peak, resides.

Baxter's Ordeal

Baxter left early morning
and swam across lake.
The rut was upon him,
'twas time to find a mate.
He swam over lake
and stood upon rocks.
then,
slipped in the water
over antler's water topped.
He stood once again,
then shook his hind den.
Then his massive head shook,
antlers shook, shook again.

Once in a while, you need to take a short break from your hiking. These provided the perfect spot for those moments of rest, and would give you time to look around yourself at your surroundings.

Sittin' Rocks

There's some good ones out there along that trail, as plain as day can be. They're called sittin' rocks; sittin' rocks, as pretty as can be. You're tired of thinking about woodland lore. Your feet get tired and bones get sore. There's some good ones out there on that trail, they're called sittin' rocks, sittin' rocks.

The Spot

There's this spot in the pines in the north woods, you know. I've been there but once, but back I would go.

The land sits in a lake on peninsula bound. There's nothing but water, and more water around.

You'll hear waves a-lapping on shore as you sit, and the soft breeze in those pines just will not quit.

A reminder that fall, that fall is nearby, for those skies begin clouding with gloom in their eye.

There's no one nearby to enjoy this with me, so I write it on paper, my thoughts that they be.

There's a spot in those pines, far north woods you know.

I've been there but once,
but
back I would go.

The White House Landing

You blow an air horn cross the lake, and Bill, he'll pick you up.

There's never any hurry, unless you want to make that sup.

I heard that Linda's burgers are quite the size, indeed.
Those pounders fill you up and . . . that is all you need.

The lake was oh so quiet, very comforting at night. Then
Linda came around, said did we care to see a sight? So . . .
we went around their cabin to see about the sight.

And . . .

There stood huge old Baxter, in all of Baxter's might. The
guys, they took his picture. He didn't seem to mind. For
food was all that mattered; that mattered in his mind.

She said he's seven years old now, and quite the sight to
see, as I watched that awesome creature in his reality.

The North Woods

Maine's north woods are so lovely, they're a beauty to behold.
Those north woods are so lovely, they're more beautiful than gold.
God made them oh so lovely for you and me to see!
For those north woods are so lovely, and that they'll always be.

The Loon

The loon's mournful call 'round midnight wakes me from my sleep.

I hear his mournful call, the call he likes to keep.

I know his call comes somewhere out on waters deep.

It's that mournful call that's out there . . . Oh the loon, he likes to keep.

Maine Sky

The night is clear and the sky is bright, for I'm under a Maine sky tonight.

As I watch that sky so brightly lit, I see not one shooting star but more that be.

I count the first, then two, three, and four.

I watch quietly, watch fascinated, watch quietly for more,

And . . .

The Big Dipper's out for me to see. She hangs in the sky in majesty.

The night is clear and the sky is bright, for I'm under a Maine sky tonight.

GG was the trail name of a woman friend that I hiked with for a short while. She hiked with a big German shepherd called Bruno. GG's trail name stood for Got Gumption, or Gumptious Gal, whichever way you wanted to take it.

GG and Bruno

GG's got this friend called Bruno. He's quite a friend, indeed.
For Bruno is the friend that watches GG's need.
He's all so all attentive and very good for her.
She depends on this friend called Bruno, and
On him she can be sure.

Understanding

Maine's such a different place for me, the beauties that unfold.
The people of Maine call a lake a pond, I just don't understand.
The people of Maine call a valley a notch. I just don't understand.
They have the strangest creatures here, I just don't understand.
The moose are huge and wondrous; those antlers big and strong.
The porcupines, they bide their time across the path, you see,
and slowly, oh so slowly, they'll climb up yonder tree.
The loons are out at night. They throw in harmony.
Horned owls come out, also, asking who. Just who's around.
In daylight hours, woodpeckers as big as crows abound.
Hunting bear is easily done, they bait them here, you know.
It's not hard to figure out, Maine puts on quite a show!

What I would call a good-sized lake in the state of Michigan (where I reside) is commonly called a pond in the state of Maine. In the nicer weather, some hikers enjoy hiking under a full moon because your night vision is so much better. One evening under a full moon in the state of Maine, I had set up my tent near a good-sized pond as the rain fell lightly. Here is how I remembered it:

Moon Rain

The rain falls . . . a silent meteor shower leaving circles
like craters
in
pools of the moon.

On (or end of) the Trail

The Appalachian Trail is roughly 2,180 miles in length.
It travels through fourteen states along the crests and
valleys of the Appalachian Mountain Range from the
Springer Mountain, Georgia, to Katadhin, Maine.

Mt. Katadhin is the highest peak in the state of Maine. If you look at a profile of it on a map, it basically shows five miles straight up.

Mounting Katahdin

As I've mounted this here boulder, it's euphoria I feel.
As I've climbed this Mt. Katahdin, it's so vast, so huge, unreal.
As I've mounted Mt. Katahdin, it's euphoria I feel.

About the Author

Stephen Lont was a Sergeant in the 82nd ABN during the Vietnam era. He has always enjoyed being outdoors and working with nature. Steve now resided in Holland, Michigan, with his family, dog (Hope), four hens, and thousands of honeybees.